ONE SMALL SQUARE®

TROPICAL RAIN FOREST

by Donald M. Silver

illustrated by Patricia J. Wynne

LEARNING TRIANGLE PRESS

Connecting kids, parents, and teachers through learning

An imprint of McGraw-Hill

New York San Francisco Washington, D.C. Auckland Bogotá
Caracas Lisbon London Madrid Mexico City Milan
Montreal New Delhi San Juan Singapore
Sydney Tokyo Toronto

Every living thing pictured in this book can be found with its name on pages 38-43. If you come to a word you don't know or can't pronounce, look for it on pages 44-47.

For Viola Marcus

our enigmatic friend.

We wish to thank Dr. Nancy Simmons of the American Museum of Natural History for her insights and suggestions. It has been a delight working with Judith Terrill-Breuer, Jane Palmieri, and Roger Kasunic on our One Small Square series. And we have deeply appreciated the artistic endeavors of Dianne Ettl and Ivy Sky Rutzky and the encouragement extended by Marc Gave, Maceo Mitchell, and Thomas L. Cathey. Special thanks to our youngest editor, McCall Elizabeth Breuer, for being so perceptive.

Library of Congress Cataloging Number applied for.
ISBN 0-07-058051-0
9 10 11 12 QTN/QTN 13

Whether you are in a rain forest or at home, always obey safety rules! Neither the publisher nor the author is liable for any damage that may be caused or any injury sustained as a result of doing any of the activities in this book.

Actual size

Introduction

Call it a piggyback ride. Call it leap frog. Call it hide and seek. Just don't call what the dart-poison frog does a game. She has no time to play when it comes to her tadpoles.

First she laid her eggs on a fallen leaf. Then she waited for the tadpoles to hatch. Next she let each one wriggle onto her back. When they were all aboard, she hopped over to a giant tree and leaped onto its bark.

Up, up, she climbed, carrying her young. Her bright colors warned frog eaters to stay away from her deadly poison. Up, up, with still no branch in sight. Up, up, without a single tadpole sliding off her back.

4

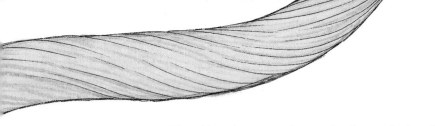

Finally the tree branched, and the frog went out on a limb. Not just any limb. This mother frog chose one where a number of pineapplelike plants were growing. Then she hid one tadpole—only one—in each plant.

The frog's work was still not done. Every few days she returned to the same plants. In each one she laid eggs full of food. Without this food her tadpoles would not grow and turn into frogs.

Why didn't the frog lay her eggs in a pond? Why did she climb such a tall tree and drop her tadpoles into plants? And why were those plants growing high up on trees instead of on the ground?

Why? Because the frog, the tree, and the plants all live in a tropical rain forest. And life in a tropical rain forest is unlike life anywhere else. More kinds of plants and animals live in rain forests than in any other habitats. And there may be millions more to be discovered.

Monera   Protists

Funguses   Animals

Plants

With the simple equipment shown here, you can do the activities in this book at home or in a nearby park. These activities will help you understand how a tropical rain forest works.

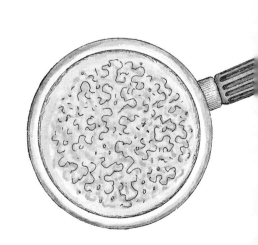

One Small Square of Rain Forest Floor

Is there a tropical rain forest close to your house? Not unless you live near the equator, where it is warm all year long and it rains a lot. But don't let that keep you from discovering the ins and outs of life in a rain forest. Then you will understand why tropical rain forests are in such danger of disappearing and why we must save them.

Climb aboard and let One Small Square transport you to a tropical rain forest. You might find such a forest in Central or South America. You can start here with a small square of the forest floor about the size of a four-person elevator—4 feet (1.2 m) on each side. Or you can start at the top of a giant tree (page 22 or page 30). Or in between (page 14).

No matter where you start, the wonders of each layer of the rain forest will unfold before your very eyes. Be on the lookout for a super-slow sloth, a wide-winged moth, a hovering hummingbird, and bats snoozing beneath a leaf. At the same time beware of boas, bloodsuckers, and biting ants. There are creatures you've never heard of, with more places to hide than you can imagine.

Perhaps most wondrous of all are the many different ways that creatures use to stay alive in the lush greenness of a tropical rain forest.

Just one small square won't do if you want to find out how a tropical rain forest works. First explore the forest floor. Next look up into the world of the understory. Then climb a tall tree to reach the canopy. And finally hold on tight as you enter the emergent layer of the forest.

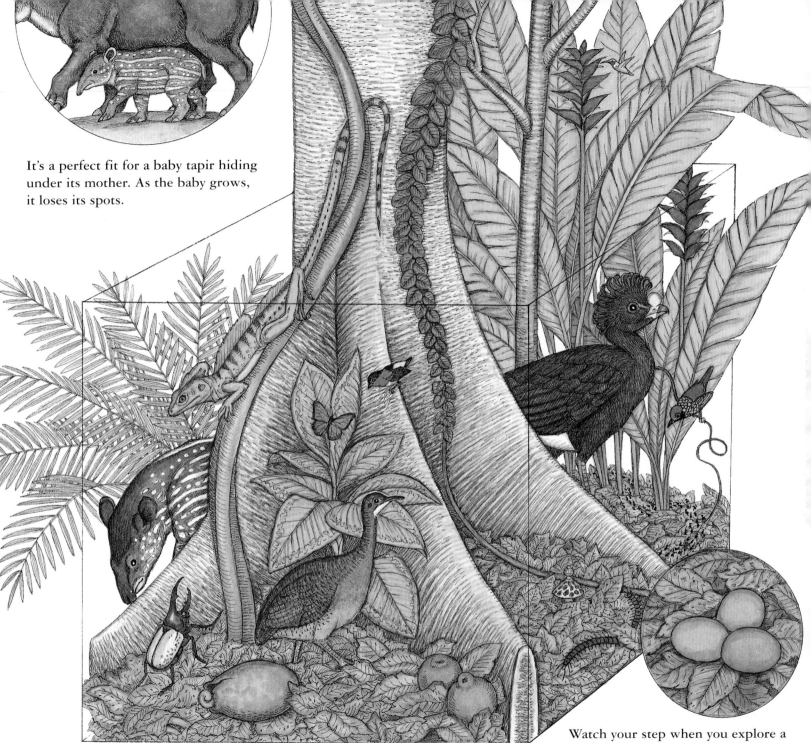

It's a perfect fit for a baby tapir hiding under its mother. As the baby grows, it loses its spots.

Watch your step when you explore a tropical rain forest. You don't want to hear a crunching sound and find out you stepped on bird eggs.

Up, up, shoots the trunk of the kapok, the giant tree in the small square. Out, out, spreads the trunk bottom. Special buttress roots grow out of it. They help prop up the tree so it does not topple.

Forget the Umbrella

When you enter a rain forest, prepare to get wet. Even before any rain can soak you, your sweat will. By the time you reach the small square, your clothes will be sticking to your body. Don't expect to dry off. In the rain forest there is no escape from heat or dampness.

Everywhere you look there are trees, trees, and more trees. Most of them could tower above a fifteen-story building. These forest giants rise straight, then seem to open like umbrellas, with their leaves and branches near the top.

How many leaves are there high up in the trees? So many that you can barely glimpse the sky. There are enough to keep the heat inside the forest and shut out breezes. And more than enough to block just about all of the sunlight from shining on you below. Think of your room when you have only a dim light on. That's how poorly lit most of the forest floor is. It's a wonder that any plants can grow there.

For plants to stay alive, their leaves need light to make food out of water and the gas called carbon dioxide. While tall trees head skyward for sun, only shade-loving plants can remain behind. They don't need as much sunlight to make food. Step into the rain forest and you will see how few shade lovers there are. That's why you can walk just about anywhere you want. If the forest floor were bathed in sunlight, it would turn into a jungle, overgrown with plants blocking your path.

It's woody. It rises to the top of the forest. No, it's not a tree but a liana vine. And if you don't watch out, it will trip you up. Lianas start to grow on the forest floor. But they will try to climb any stem or trunk they touch.

Like a rug, a mat of roots covers parts of the forest floor. If you look at a root under a magnifying glass, you will see smaller roots coming out of it.

9

The Ants Go Marching

In your backyard, when the ants go marching one by one, do you even notice? At a picnic, when they march two by two over your food, maybe they catch your eye. But in the rain forest, when the ants go marching by the tens of thousands, you get out of their way! Otherwise you will be bitten by the large soldiers' hooked jaws or stung by their tails. That's right, soldiers. An army of ten million ants roams this part of the forest.

Army ants are predators, animals that eat other animals for food. They move from place to place, set up camp, and send scouts in search of prey. When one scout comes upon the small square, it hurries back to the others, leaving an odor trail behind. Soon the army is marching along the trail, off to raid the square.

If an antbird gets in the army's way, the ants will attack it like everything else. Why don't the birds eat the ants? Because a meal of army ants makes birds sick.

Not even the scorpion's deadly sting can save it from army ants. They will carry off their prey and feast on it later.

10

The giant Hercules beetle dines on plant juices, but its young eat rotting leaves and wood. Along with snails, velvet worms, and termites, they are part of the forest cleanup crew.

All is quiet. Then, without warning, the ants appear. They climb leaves, invade holes, and attack every creature in their path. Beetles, roaches, spiders, and lizards try to flee to safety, often to be snatched up by antbirds that follow the army.

When the raid is over, the ants march off, carrying a captured scorpion. Although for a brief time they turned the small square into a battlefield, they did not leave it lifeless. Out of the earth crawl worms, centipedes, and other ants. A wasp that escaped flies back into the square. And many animals the ants did not reach stay where they are hidden.

With hooked legs, army ants link up like bricks to form a living wall. Behind the wall the army rests for the night inside a rotting log.

Your Rain Forest Notebook

One day you may get to explore a tropical rain forest. Until then, you can keep a rain forest notebook. In it use words and pictures to record the activities from this book that you do. Paste in newspaper reports and magazine photos about tropical rain forests. If you ever do get the chance to explore a forest, bring your notebook along.

A Square to Start With

Your backyard is no rain forest. Nor is the town park. But in both places you will find lots of life above and under the ground. Explore one small square of your backyard or park. What kinds of plants grow there? Where are their roots? How many ants, beetles, and other insects can you find? Look at them under a magnifying glass and draw what you see in your notebook.

Do you think it's easy to find camouflaged animals? How many can you uncover on these two pages? Which was the hardest to find?

Where are they? Right in front of your eyes. Just don't expect to see them. These hiders are hard to spot because their shapes, colors, and patterns blend into the world around them. This ability, called camouflage, can be a life saver. If you overlook these creatures, so might a predator. Indeed, only when a stick walks, a thorn hops, a seed leaps, or a leaf flaps its wings do these animals give themselves away.

There's something else going on that is easy to miss. An army of recyclers—bacteria, protists, funguses—is attacking fallen leaves and twigs, dead animals, even droppings. Lift a few fallen leaves. Do you see a wispy web underneath? It is made of fungus threads that invade

A leaf butterfly is quite a fooler. All it has to do when it lands is open its wings and stay still. Then it looks just like a leaf. Can you find two of these insects on this page?

For a lantern fly, two tricks are better than one. If blending into bark doesn't fool a predator, it opens its wings, flashing its two eye spots. These confuse the predator, and the fly can zoom away.

the leaves, break them down, and soak up their nutrients. The funguses use some of the nutrients for food. But the threads deliver other nutrients, such as minerals, to plant roots on the forest floor. The minerals don't wind up in the soil, where rain could wash them away.

The recyclers work day and night all year long. Everything in the rain forest depends on them. Without minerals, trees could not grow. Neither could other plants. The forest would slowly die off. There would be no home for hunters or hiders. And no place for the ants to go marching.

Wherever the clear wing flies, it looks like something else. That's because, as this butterfly's name suggests, you can see right through its wings. Something else on this page is see-through. Do you know what?

The Layered Look

If you take away all the tallest trees in the rain forest, what is left? A forest of shorter trees that might remind you of the woods closer to home.

The next time you visit the woods or a park with lots of trees, see what is growing there.

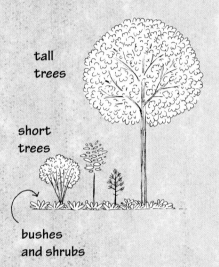

tall trees

short trees

bushes and shrubs

How many different plants can you find? Which grow in bright sunlight? Which love the shade? Is there a layer of ground-hugging plants in your woods? How about bushes and shrubs? Do short trees grow above the bushes, and taller trees above them?

In your notebook draw all the different layers and label them. If you see any animals, draw them in. Use a magnifying glass to search for insects and other creatures on leaves and bark.

One Small Square of the Understory

Have you ever looked up a tree? If not, try it. Unless all its leaves are gone, you will have to work hard to find animals on branches high above the ground. A flash of color may tip you off that a bird or a butterfly has landed. Without binoculars, you may be unable to catch more than a fleeting glimpse of the animal.

Spotting animals up a rain forest tree is even harder. The light is dim, and leaves and vines are everywhere. You would have to take away the giant trees to get a good look at the shorter ones. These are the trees that slowly grow in the understory—the layer of the rain forest sandwiched between the ground and the tops of the giants.

Let's explore one small square of the understory. If you come across a jaguar asleep on a branch, be sure not to startle it!

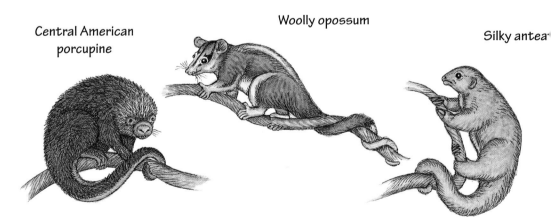

Central American porcupine

Woolly opossum

Silky antea[r]

Curved claws sure come in handy for climbing trees and vines. But so does a wraparound tail. It is like an extra leg that can grasp branches or hook onto them for just hanging around.

14

Black-throated trogon

Motmot

In the rain forest, wherever there are insects, there are insect eaters. When this motmot catches a butterfly, it beats the wings off before eating its prey. The trogon catches insects in midflight.

A hooklike claw can't be beat if you are a sloth that spends the day hanging upside down.

Liana

Up is the way to go in search of a termite nest the size of a soccer ball. Inside are tasty termites, a tamandua's favorite food. But dinner won't be all fun. Termite soldiers will spray a tamandua with sticky fluid, trying to chase it away.

16

The Way to Go

Something is moving down one of the vines. It sure looks like a parade of leaves. But it can't be. Leaves have no legs. Maybe it is an animal that looks like a leaf. Get closer. It's a parade of ants! Each ant is carrying a piece of a leaf on its back. Some of the pieces are bigger than the ants holding them.

Follow those ants. It's O.K. They're not army ants but leafcutters. They will lead you to their underground nest. Too bad you can't morph into one of them. If you could, you would be met by an ant guard at the nest opening. The guard would smell you with its antennas to make sure you belong in the nest.

Once inside, you would make your way through tunnels until you reached a chamber. Other ants are there, chewing pieces of leaves into a mushy ball. They add each new ball to a mound of mush that has fungus growing on it.

Can you guess why the ants do this? It's because they are gardeners. The mushy mound is their underground fungus garden. The fungus is the only food the ants eat. They grow it on the leaves they slice up with their sawlike jaws.

Backtrack to the vine and watch the leafcutters come and go. Look for other creatures too. In the rain forest, vines and trunks are highways from the ground to the tops of trees. They are the way to go if an animal can't fly.

Weight lifters can lift and carry something bigger than they are. So can leafcutter ants. But they can climb down a vine at the same time!

Growing a garden is hard work. Not only the leafcutters' funguses grow in it but so do other funguses. The ants weed them out.

The small ant on the leafcutter's leaf is no hitchhiker. It is riding shotgun. It will attack any fly that tries to lay eggs on the back of the ant carrying the leaf.

Let the Sun Shine In

Who turned on the lights? In a corner of the small square, a broad beam of sunlight shines through the roof of leaves that covers the rain forest. All it took was a storm in which a tall tree fell and knocked over other trees nearby.

Almost at once, bacteria and funguses started to grow on the fallen trees. Termites also moved in. With the help of protists in their stomach, the termites are able to eat wood and digest it.

Something else happened. When sunlight hit a slow growing short tree, that tree started to shoot up quickly. Unless another tree outgrows it, it will take the place of the one that fell.

Colorful passionflowers blossomed. And with them came their enemies, Heliconid butterflies. Heliconids lay eggs on the

When the Heliconid caterpillar turns into a butterfly, what happens to the poison inside? It stays there. The butterfly's bright colors warn birds that this is one meal to avoid.

Bees, birds, and butterflies can't resist the passionflower's sweet nectar.

This hummingbird makes it look easy to hover in midair while sipping flower nectar.

Call this plant "ants' delight," but it is really an acacia tree. The tree grows hollow thorns where ants make their home. It also drips drops of nectar the ants love. In return, the ants attack leaf eaters and chew away any vine that tries to climb their tree.

leaves of passionflower vines. When the eggs hatch, the caterpillars eat the vines' leaves. Few other insects do because the leaves are full of poison. Caterpillars are not only unharmed by the poison, but they store it inside them. Their bright colors warn predators that a bite of caterpillar means a bite of poison too.

What stops the caterpillars from eating all the leaves? The vines do. Some grow hooks that stab caterpillars. Others make a sweet liquid that attracts ants. The ants, in turn, attack any caterpillar trying to harm the vine that feeds them.

Slow-mo Sloth-mo

A sloth spends most of its life hanging upside down from a tree. When it does move, it never goes fast. If it covers 15 feet (4.5 m) in one hour, it is close to breaking the sloth speed limit.

Measure off 15 feet (4.5 m) in your home, backyard, or park. How long does it take you to walk that distance? Try moving so slowly that it takes you an hour. Moving super slowly may get on your nerves, but it helps a sloth stay alive. The sloth gets where it is going without jaguars and other sloth eaters noticing that it is even moving.

15 feet

tape measure

Anting

You don't have to visit a rain forest to find leafcutter ants. Some kinds live as far north as New Jersey. Many zoos and conservation parks display leafcutters, nest and all. Look for ants in your yard or park. Are any carrying bits of food that people dropped? Can you find the opening to their nest?

The squeak of a mouse is all the crested owl needs to hear in order to find its prey in the dark. No wonder the curassow stays quiet with a jaguar so close by.

The Night of the Vampire

Darkness falls on the rain forest. Owls hoot and monkeys howl. Spiders and scorpions are on the prowl. A jaguar awakes hungry, its stomach growling. Nearby a cat-eyed snake uses its forked tongue to taste what will be a feast of frog eggs. It is feeding time for the creatures of the night.

There are hunters in the trees, on the vines, and in the air. They see, feel, smell, and hear their way to food. A long-nosed bat flies after moths and other insects. A little white bat sniffs the air for ripe fruit. And if a vampire bat finds a tapir asleep on the ground,

Even with a mouthful of fruit, this bat still sends out very high-pitched sounds from its nose. They bounce off vines and branches, and back to the bat, telling it where not to fly.

Too bad for the red-eyed tree frog. A big spider is about to grab and eat it.

it will use its sharp teeth to slice into the tapir's skin. After the bat laps up a meal of warm tapir blood, it will zoom off without ever waking the mammal.

Jaguars are the largest predators in the understory. But you don't want to tangle with the smaller predators, either. Spiders and scorpions prey on frogs, lizards, small birds, and even poisonous snakes. Every rain forest creature must eat to stay alive, so every creature is also at risk of being eaten.

Not all night animals are out for food. While a male frog was calling for a mate, a female frog laid eggs. She sure picked a bad spot. Where are her eggs about to wind up?

The tayra chasing the tree rat is expert at running along vines without falling. But it better be careful that some other animal doesn't catch it.

Rain isn't the only thing falling in a tropical rain forest. Leaves, flowers, air plants, fruits, seeds, sometimes even animals, tumble out of the canopy and down to the ground. They are clues to what is living in the part of the forest that is difficult for people to reach. To get a close-up view of the canopy, you would need rock-climbing ropes, a tall crane with a basket, or a balloon raft that could rest on treetops. So if you ever do go to a tropical rain forest, most likely you'll have to be content with what you find on the ground.

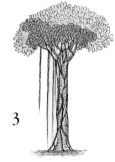

Piece of paper

One Small Square of Canopy

Exploring a small square of the canopy is like discovering a new world. Creatures live there that never come down to the ground. Like giant snakes, tangles of vines loop from branch to branch and from tree to tree. Everywhere there are leaves. And there are hanging gardens of air plants.

Air plants, or epiphytes, grow on trunks, branches, and vines. Their roots never touch ground. Each air plant gets the water and minerals it needs to stay alive and make food from air, rain, and decaying leaves and animals.

If you see roots dangling from a high branch all the way down to the soil, they may belong to the deadly strangler fig. Don't worry: The strangler won't attack you—unless you are a tree. A strangler starts life as an air plant on a tall tree's branch (1). Then it sends out roots that reach the ground and soak up water and minerals (2). As the strangler grows, it branches and sends down more roots. But these roots wind around the tree's branches and trunk (3). They form a cage with the tree trapped inside.

The tree is a goner. From above, the strangler's leaves block its light. From below, the strangler's roots steal water and minerals the tree needs. As the trunk tries to grow, the cage of roots strangles it. Slowly the tree rots. But the strangler is left standing in the same spot (4). What some plants won't do for a place in the sun!

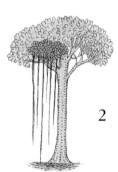

3

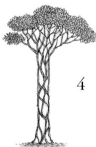

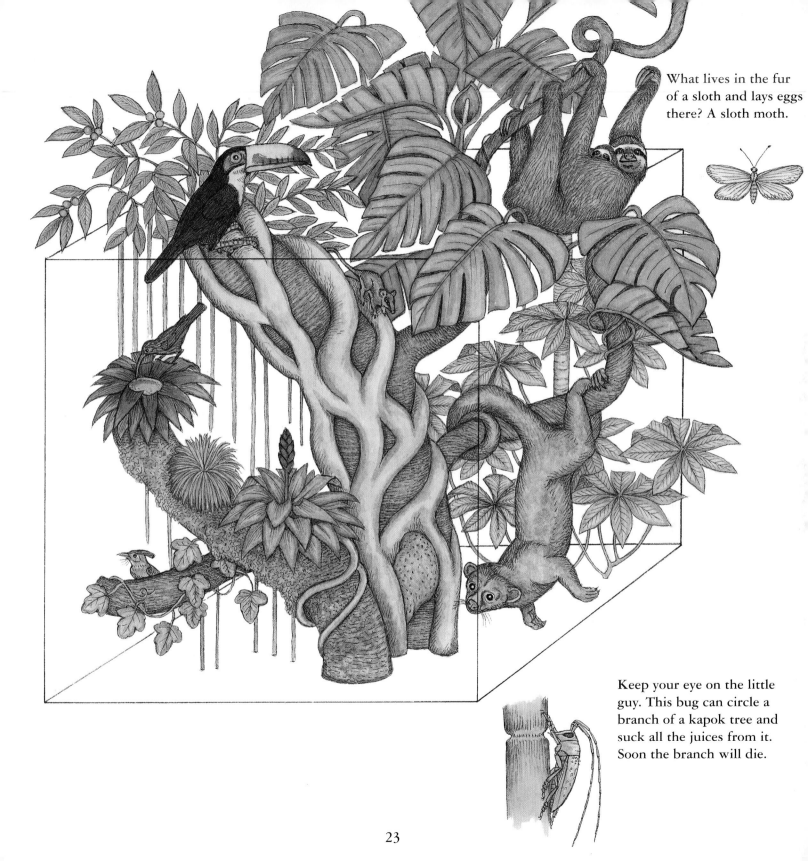

What lives in the fur of a sloth and lays eggs there? A sloth moth.

Keep your eye on the little guy. This bug can circle a branch of a kapok tree and suck all the juices from it. Soon the branch will die.

Sunlight

Chloroplasts

Countless leaves make food in the canopy. Inside each leaf are chloroplasts that soak up the energy in sunlight needed to make the food.

Drip tip

Want rain to roll off instead of hanging around? Take a tip—from the drip tip, that is. You'll find it on leaves that narrow and point down.

Up in the Treetops

What's the "buzz"? Up in the treetops, flowers are bursting into bloom. Each flower that opens calls attention to itself with colors or scents—or both. It is advertising its own brand of sweet nectar and energy-rich pollen. No wonder the bees are buzzing. In fact, the small square up in the air is humming with activity. Lots of hungry animals want to join the feast: birds, beetles, flies, butterflies, and wasps.

A tropical rain forest is the perfect place for nectar and pollen feeders to live. There is no autumn, with trees losing their leaves. Instead, most trees and vines just lose older leaves and grow new ones. The canopy is always a sea of

Insects beware: Not everything that looks like a flower *is* a flower. One of these is a crab spider that will eat you if you come too near.

Crab spider

When the sun goes down, some flowers open for business. They serve sweet nectar to night-flying moths and bats.

green. And different plants can bloom at different times, so there is a supply of flowers year round.

A flower's job is to make seeds so new plants can grow. To do that, pollen from male flower parts must be spread to female parts on the same kind of flower. While some plants depend on wind to spread pollen for them, most enlist the help of bees and other nectar and pollen feeders.

When a bee, for instance, lands on a flower, it tries to reach the nectar. That's when some flower pollen sticks to it. At the next flower the pollen may drop onto or rub against a female part. The bee gets fed, pollen gets spread, and the forest gets new seeds.

This tanager came for a taste of nectar and will leave carrying sticky pollen.

The bucket orchid takes no chances getting pollinated. When a bee lands, it falls into the "bucket" of water. The only way out is to climb up to a narrow opening. There the flower's pollen sticks to the bee's wet back.

A bromeliad takes up branch space, but it does not harm a tree.

Bacteria and protists break down bits of leaves and dead insects that fall into the water. They free nutrients that the bromeliad needs to keep growing.

Even up here there are ants. They live inside the bromeliad's stem.

Can you find the little hummingbird sitting in her nest? She's only an inch (2.5 cm) long. Her eggs are the size of peas.

Fill That Tank

Believe it or not: In spite of all the rain, the tops of rain forest trees can dry out after a downpour. Water drips off leaves and streams down branches. In the heat of the sun, drops that remain evaporate—turn into gas. Yet there must be water somewhere, since so many creatures live in the canopy.

There is. The leaves of the bromeliad plant overlap at the bottom and press together. They form a tank that can hold as much as a gallon (4 l) of rain. This water attracts animals, and soon the tank bustles with life. It is like a minipond in the sky. Over 250 kinds of creatures may depend on it to help them stay alive.

One minipond is a bath for birds, a watering hole for lizards, and sometimes a bathroom for both. Beetles and crabs live in it. So do young mosquitoes and dragonflies. A cricket may jump in to escape being eaten. Snakes, monkeys, and opossums find it a sure spot in the treetop for a snack. And for a tadpole dropped in by its mother, the minipond is the place to lose its tail, grow legs, and turn into a grown-up dart-poison frog.

Is a place in the sun worth the climb to the top? It is for vines. The canopy is where their leaves can make food and their flowers bloom.

Eat Me

Most fruits have to be eaten for the seeds inside to be spread by animals. The fruits attract fruit eaters with their colors, odors, and sweet tastes. Do any trees in your backyard or in the park bear fruit? If so, which animals feed on them? When fruits drop to the ground, which creatures nibble them? Do any predators make a meal of eating the animals that eat the fruit?

Don't Eat Me

Why do so many rain forest plants grow leaves that are tough, bad-tasting, or poisonous? Why do they grow thorns and spines? To keep away as many kinds of insects and other plant eaters as they can. What harm can plant eaters do? Find out on a spring or summer day. Use your magnifying glass to look at leaves on backyard or park plants. How many have rips, tears, or bite marks? Have any plants been eaten away?

With a berry in its rainbow beak, what can a toucan do? Tilt back its head and toss the treat down its throat.

Don't let size fool you: A bell bird can raise a racket.

Pluckers and Noisemakers

If you were up here, what would catch your eye? A toucan plucking fruit with its big, colorful beak? A monkey leaping from one tree to another like an acrobat? A quetzal feeding on an avocado as its long green tail dangles like a vine? Maybe you prefer listening to the noisy squawkers, screamers, and howlers hidden by canopy leaves. Just don't miss the fruits.

Many flowers have turned into fruits full of seeds. When the fruits ripen, their skins often turn red, orange, or yellow. The fruits soften and sweeten. They become easy to find,

No road signs are needed up here. Troops of howler monkeys run along certain branches more than others. These highways in the sky lead to food and safe hiding places.

Fruit bat with fig

Who can resist a branch full of figs? Not this night-flying fruit-eating bat. It will feast on every ripe fig it can. The more it eats, the more seeds it drops as it flies.

From the ground you won't be able to see much going on in the canopy. But you will hear it. Loud sounds carry well in spite of all the leaves and branches.

pick, and eat. They're tasty too. That's how they start to get the job done of spreading the seeds inside. The rest is up to the fruit eaters.

When an animal bites into a fruit, it might cause some seeds to drop. If it swallows the seeds, that's even better. Instead of being broken down and digested, many of the seeds pass right through the animal's body. They wind up in its droppings. A bird can eat a fruit in the small square, then fly to another part of the forest before dropping the seeds. If the seeds fall where the sun shines, they may grow into a new small square.

Mealy parrot

Part of Your Life

In one way or another, tropical rain forests are part of your life. But you may not realize it. Corn and tomatoes were first found growing there. Sugar, bananas, chocolate, and spices come from rain forest plants. So do many medicines and soaps.

Do you have any houseplants that grow in the shade? They may be from a rain forest. Do colorful birds stop over in your town in autumn or spring? They may be on their way to or from a rain forest.

To find out, look in a field guide. It contains the names and pictures of animals and plants that live in different places. A field guide to birds tells you which rain forest birds you may find in your town. You may have field guides at home. If not, you can find them at the library. In your notebook, make a list of everything that comes from a rain forest and touches your life.

One Small Square of the Emergent Layer

Welcome to the attic of the rain forest—the emergent layer. It's made up of the tops of a few trees that stick up above the rest. From here the view is spectacular. Shiny bright-blue butterfly wings sparkle in the brilliant sunlight. And flashes of red signal parrots feeding and flying in the canopy below.

This small square belongs to the harpy eagle, the most powerful eagle in the world. A fierce hunter, it can swoop down like a dive bomber through the leaves to seize a monkey or a sloth with its sharp, curved claws. There is little chance for escape. As the claws tighten, the hooks stab into the prey. The eagle then returns to its platform nest in the tallest treetop with food for itself and its young.

Brown-hooded parrot

White-fronted parrot

Barred parakeet

The emergent layer is parrot country. Colorful parrots feed and nest there. A male scarlet macaw courts a female by offering her a fruit. If they mate, she will lay eggs inside her tree-hole nest.

Scarlet macaw

30

If there is a way
to reach a meal, a
snake will find it.

When a young harpy eagle is hungry,
it screams for its parents to feed it.

31

Out for a swim, a matamata will suck in every small creature that falls into the water. With so many insects around, the turtle rarely goes hungry when the forest floods.

Fishes are fun to look at. But don't miss the baby hoatzin. It has wing claws that hook on to vines.

A Special Place

Just about every day some rain falls in the forest. But certain times of the year are wetter than others. So much rain can fall then that the forest floods. And if a nearby river overflows its banks, the forest will come alive with catfishes, pacus, and matamata turtles.

Neither floods nor heavy rains harm the forest. What does harm them are people who cut down trees and vines to make room for farms. Every second of every day they are destroying a piece of rain forest the size of a football field.

Baby hoatzin

They are taking away the homes of millions of kinds of wildlife with nowhere else to go. And people often do not understand that the forest soil lacks the nutrients that crops need to grow.

That's why so many nature lovers are writing to government leaders, asking them to step in and stop the destruction. They are spreading the word that rain forests are special places full of living wonders. Some students are even raising money to buy pieces of forest. They are doing their share to try to save the rain forests one small square at a time.

The tambaqui feeds on fruit that falls to the forest floor. Fishes can help scatter seeds just like birds do.

Rain Forest in a Box

Stand a shoe box on its end. Measure the width and height of the box. Cut a piece of paper for the background wall about ⁄ inch (6 mm) shorter than the box height and about 4 inches (10 cm) longer than its width. Draw and color trees the full height of the box. You can use the picture on page 6 as a guide.

On separate sheets of paper, draw and color shorter trees, each with a flap at the bottom. Cut out each tree, bend its flap, and glue or tape it in place in front of the tall trees. Tape or glue yarn or string onto the trees as climbing vines. Draw, color, and cut out animals. Glue them in the layer where they belong and fill your rain forest with creatures.

Match Game

Here are the four small squares of tropical rain forest. Do you know in which rain forest layer each living thing belongs? Can you match each living thing to its outline?

Royal flycatcher

Puffball fungus

Kinkajou

Carapa seed

Blue morpho butterfly

Leafcutter ants

Two-toed sloth

Termite nest

Toucan

Cecropia

Gecko

Long-tailed hummingbirds

Kapok tree

Trogon

Orange-collared manakin

Emerald boas

Cotinga

Tinamou

Palm

Centipede

34

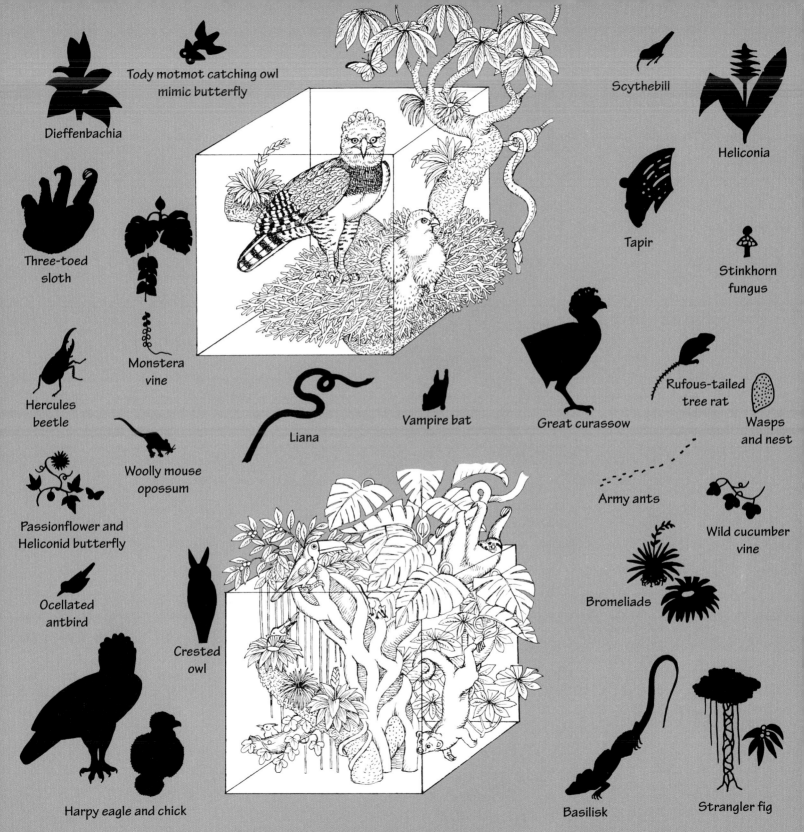

Dieffenbachia

Tody motmot catching owl mimic butterfly

Scythebill

Heliconia

Three-toed sloth

Monstera vine

Tapir

Stinkhorn fungus

Hercules beetle

Liana

Vampire bat

Great curassow

Rufous-tailed tree rat

Wasps and nest

Woolly mouse opossum

Army ants

Wild cucumber vine

Passionflower and Heliconid butterfly

Bromeliads

Ocellated antbird

Crested owl

Harpy eagle and chick

Basilisk

Strangler fig

35

The World's Tropical Rain Forests

Around the world rain forests grow just north and south of the equator. They cover only 7 percent of earth's surface, but they are home to more kinds of life than anywhere else. Countless small squares of rain forest have never been explored. Perhaps you will explore one someday, if only the world's rain forests can be saved.

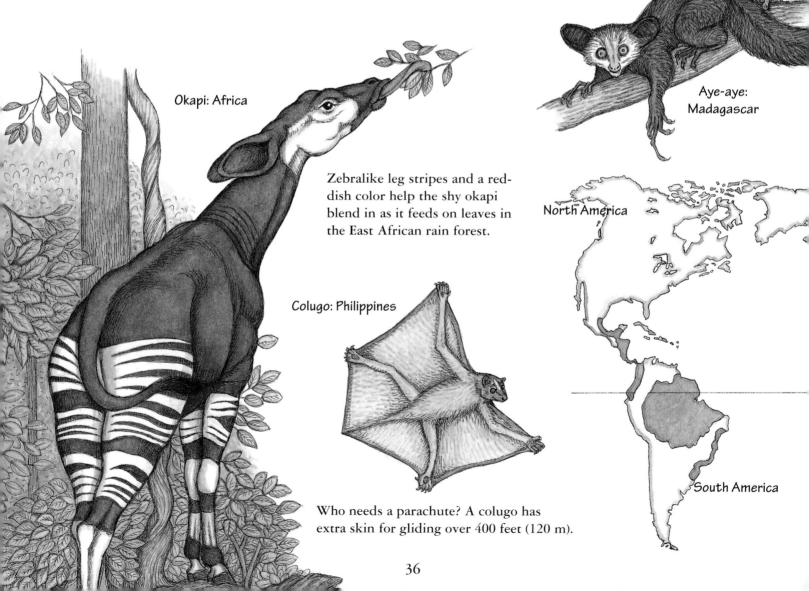

Cuban tree snail

Rain forest snails live on the forest floor and also up in trees.

Aye-aye: Madagascar

Okapi: Africa

Zebralike leg stripes and a reddish color help the shy okapi blend in as it feeds on leaves in the East African rain forest.

North America

Colugo: Philippines

South America

Who needs a parachute? A colugo has extra skin for gliding over 400 feet (120 m).

The large, powerful Bengal tiger is an endangered species in need of saving.

Bengal tiger: India

An aye-aye's long middle finger is perfect for stabbing insects out of decaying wood.

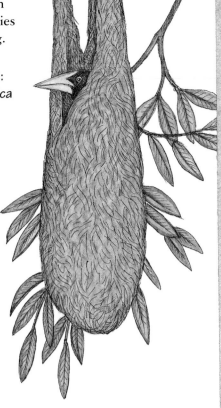

Oropendola: South America

A hanging nest keeps out most predators. A nearby nest of stinging wasps helps too.

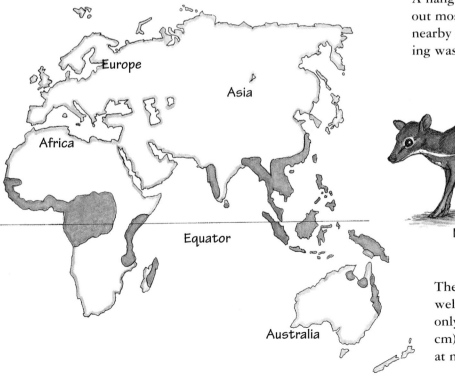

Europe

Asia

Africa

Equator

Australia

Mouse deer: Thailand

The mouse deer is well named. It is only 20 inches (50 cm) long and feeds at night.

Rain Makers

Place a plastic bag around the leaves of a small houseplant in a pot. Be sure it is tied tightly around the pot so no air can get in. How long does it take for water drops to appear on the inside of the plastic? Where do you think the water comes from?

Water rises from a plant's roots through the stem and into the leaves. There, some water is used to make food and to keep leaf cells alive. The rest escapes out of tiny openings in the leaves. It enters the air as a gas—water vapor. When the air cools, the gas turns into droplets that help form clouds. If the air keeps cooling, the droplets get bigger and bigger until they fall as rain.

The countless leaves in a tropical rain forest use lots of water to make food and stay alive in the heat. But they also give off lots into the air to be recycled into more rain. The plants help keep the rain forest rainy.

The largest creatures in the rain forest are vertebrates—animals with backbones. Mammals are the only vertebrates (and animals) that grow fur and make milk for their young. Birds are the only ones with feathers. Both mammals and birds can make their own body heat. Fishes, amphibians, and reptiles cannot make their own body heat.

Birds

Tinamou

Dotted-winged antwren

Orange-collared manakin

Great curassow

Potoo

Tody motmot

Rufous-tailed hummingbird

Black-throated trogon

Long-tailed hermit hummingbird

Royal flycatch

Ocellated antbird

Nightjar

Crested owl

Mot ow

Mammals

Agouti

Central American porcupine

White tent bat

Tapir

Woolly opossum

Tayra

Ocelot

Silky anteater

Kinkajou

Spider monkey

Brocket Deer

Three-toed sloth

Rufous tree rat

Woolly mouse opossum

Black howler monkey

Bengal tiger

Aye-aye

Mouse deer

Two-toed sloth

Tamandua

Long-tongued bat

Proboscis bat

Okapi

Vampire bat

Jaguar

Geoffroy's tamarin

Fruit bat

Colugo

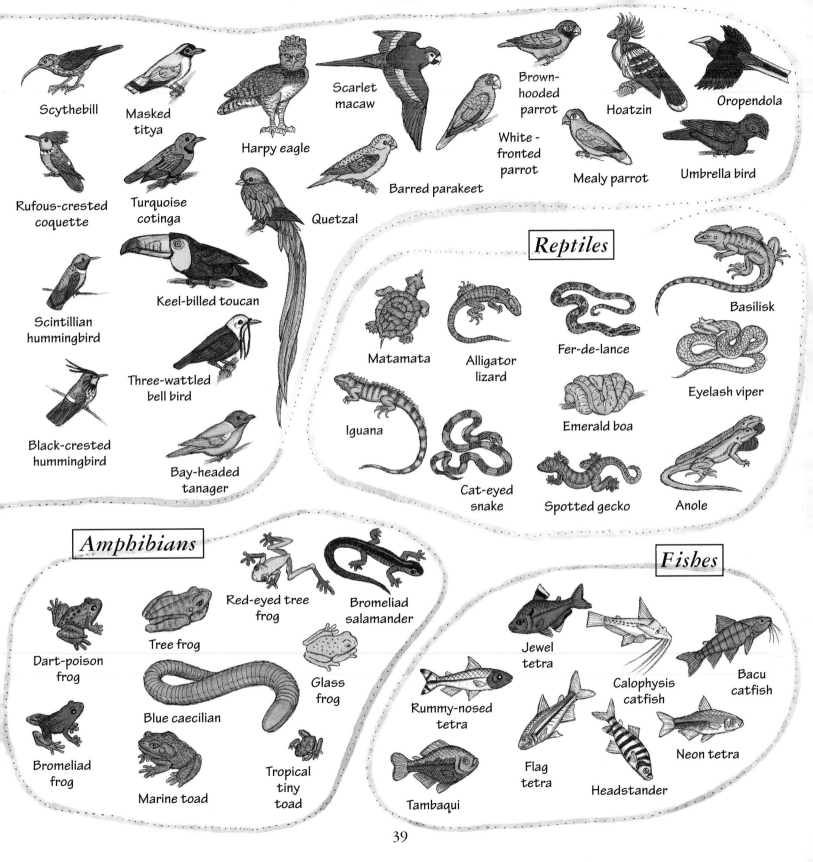

Scythebill

Masked titya

Harpy eagle

Scarlet macaw

Brown-hooded parrot

White-fronted parrot

Hoatzin

Oropendola

Rufous-crested coquette

Turquoise cotinga

Quetzal

Barred parakeet

Mealy parrot

Umbrella bird

Scintillian hummingbird

Keel-billed toucan

Three-wattled bell bird

Black-crested hummingbird

Bay-headed tanager

Reptiles

Matamata

Alligator lizard

Fer-de-lance

Basilisk

Iguana

Emerald boa

Eyelash viper

Cat-eyed snake

Spotted gecko

Anole

Amphibians

Red-eyed tree frog

Bromeliad salamander

Fishes

Dart-poison frog

Tree frog

Glass frog

Jewel tetra

Calophysis catfish

Bacu catfish

Bromeliad frog

Blue caecilian

Rummy-nosed tetra

Neon tetra

Marine toad

Tropical tiny toad

Tambaqui

Flag tetra

Headstander

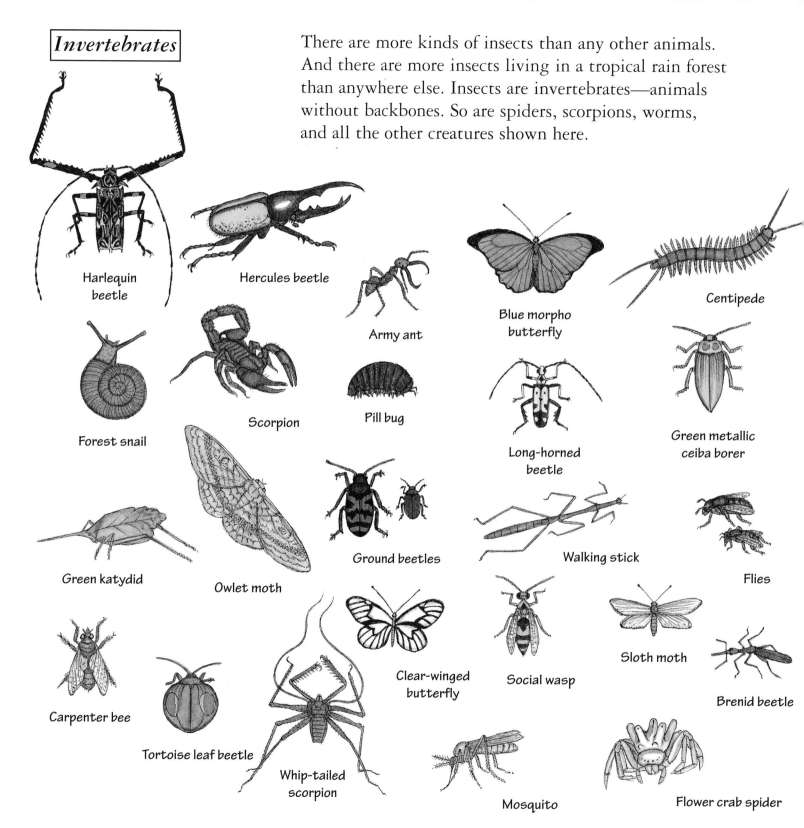

Invertebrates

There are more kinds of insects than any other animals. And there are more insects living in a tropical rain forest than anywhere else. Insects are invertebrates—animals without backbones. So are spiders, scorpions, worms, and all the other creatures shown here.

Harlequin beetle

Hercules beetle

Army ant

Blue morpho butterfly

Centipede

Forest snail

Scorpion

Pill bug

Long-horned beetle

Green metallic ceiba borer

Green katydid

Owlet moth

Ground beetles

Walking stick

Flies

Carpenter bee

Tortoise leaf beetle

Whip-tailed scorpion

Clear-winged butterfly

Social wasp

Sloth moth

Brenid beetle

Mosquito

Flower crab spider

40

Fungus-eating
beetle

Golden beetle

Velvet worm

Flatworm

Heliconid butterflies

Dead-leaf moth

Giant millipede

False-headed
caterpillar

Lanternfly

Ogre-faced
spider

Ceiba girder

Owl mimic
butterfly

Termites

Tree ants

Leafcutter ant

Scarab beetle

Bromeliad crab

Bromeliad
cricket

Bromeliad butterfly
and larva

Bromeliad snail

Thorn bug

Click beetle

Jewel bug

Fig wasp

Cuban tree snail

Frog-eating spider

Without trees there would be no rain forest. Trees and other plants are nature's food makers. Funguses often look like plants, but they can't make their own food.

Plants

Carapa seed

Kapok (Ceiba) tree

Heliconia

Monstera vine

Strangler fig

Liana vine

Palm

Pepperomia vine

Voyria

Stilt tree

Mistletoe

Wild cucumber vine

Cordia tree

Bucket orchid

Sandhopea orchid

Bromeliad

Lesica orchid

Tropical nutmeg

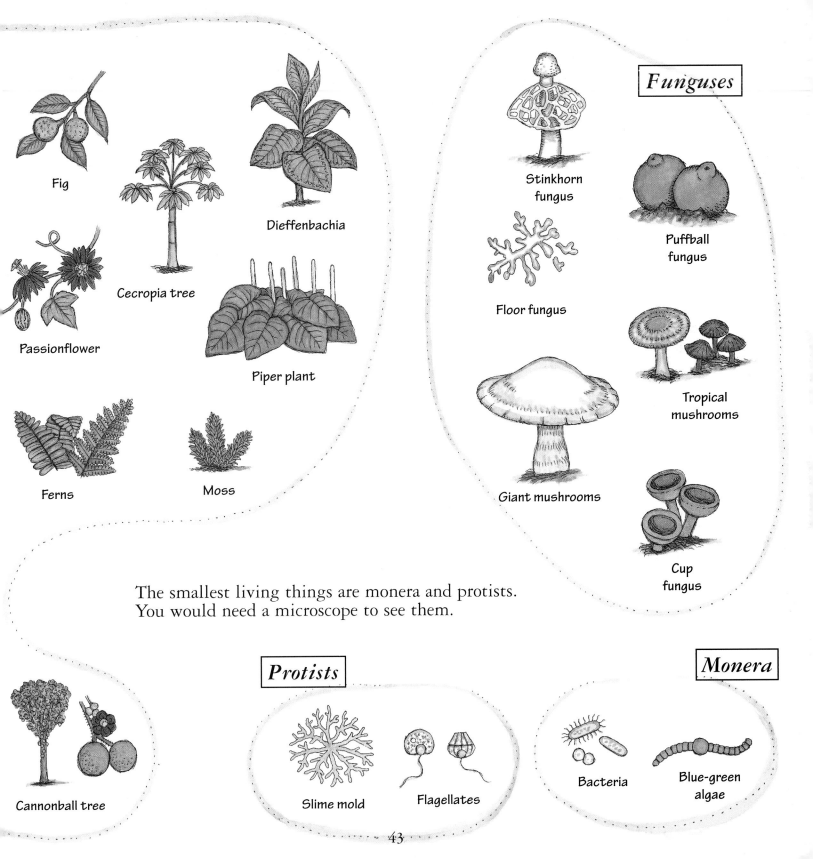

Fig

Dieffenbachia

Cecropia tree

Passionflower

Piper plant

Ferns

Moss

Funguses

Stinkhorn fungus

Puffball fungus

Floor fungus

Tropical mushrooms

Giant mushrooms

Cup fungus

The smallest living things are monera and protists.
You would need a microscope to see them.

Protists

Monera

Cannonball tree

Slime mold

Flagellates

Bacteria

Blue-green algae

43

Index

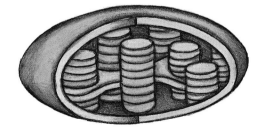

Index

chocolate 30

claw 11, 14, 16, 30, 32
cloud 37
color 12, 24, 28
colugo (kuh-LOO-goe) 36
corn 30
crab 27
crab spider 24
cricket 27
curassow (KURE-uh-soe) 20

D
dampness 9
dart-poison frog 3, 4, 27

deer 37

digestion 18, 29. *The process of breaking food down into simple parts called nutrients that an animal's body needs for growth or as an energy source.*
dragonfly 27
drip tip 24
dropping 12, 29

E
eagle 30
egg 3, 4, 7, 17, 19, 21, 27, 30. *To scientists, an egg is a female reproductive cell in a plant or an animal.*

emergent layer 6, 30, 31
endangered spicies 37
energy 24. *Ability to do work or to cause change in matter.*
epiphyte (EP-uh-fite) 22
equator 6, 36
evaporation (uh-vap-uh-RAY-shin) 27. *The process of changing from a liquid into a gas.*
eyespot 13

F
farm 32

feather 38

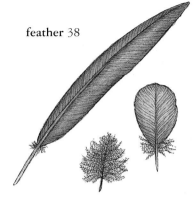

field guide 30
fig 22, 29
finger 37
fish 32, 33, 38
flood 32
flower 19, 22, 24, 25, 27, 28
fly 13, 17, 24
food 4, 9, 16, 20, 22, 24, 27, 29, 30, 37, 41
forest floor 6-13, 33, 36
frog 3, 4, 21
fruit 20, 22, 28, 29, 30, 33. *The part of a flowering plant that contains the seeds.*

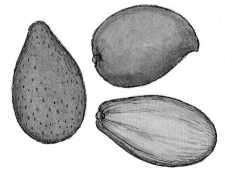

fungus 5, 12, 13, 17, 18, 43
fur 23, 38

Index

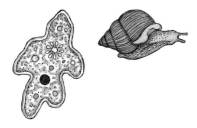

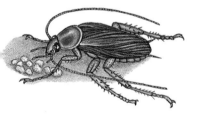

Further Reading

To find out more, look for the following in a library or bookstore:

Rain Forests: Tropical Treasures by National Wildlife Federation, Learning Triangle Press, New York, NY, 1997

Rain Forest by Robin Bernard, Scholastic, New York, NY, 1997

Why Save the Rain Forest? by Donald Silver, Julian Messner, New York, NY, 1993

Golden Guides, Golden Press, New York, NY

Golden Field Guides, Golden Press, New York, NY

The Audubon Society Beginner Field Guides, Random House, New York, NY

The Audubon Society Field Guides, Alfred A. Knopf, New York, NY

The Peterson Field Guides, Houghton Mifflin Co., Boston, MA

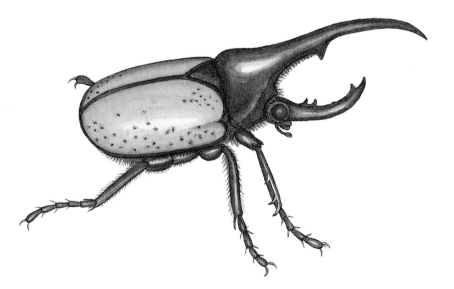